Benjamin Eddy Cotting

Disease

A Part of the Plan of Creation

Benjamin Eddy Cotting

Disease
A Part of the Plan of Creation

ISBN/EAN: 9783337816476

Printed in Europe, USA, Canada, Australia, Japan

Cover: Foto ©berggeist007 / pixelio.de

More available books at **www.hansebooks.com**

THE

ANNUAL DISCOURSE

BEFORE THE

`MASSACHUSETTS MEDICAL SOCIETY,

MAY 31, 1865.

BY BENJAMIN E. COTTING, M. D.

"A mighty maze! but not without a Plan."

BOSTON:
DAVID CLAPP & SON—334 WASHINGTON STREET.

1866.

Re-printed from the

"Medical Communications of the Massachusetts Medical Society,"

Vol. X.......No. V.......1865.

David Clapp & Son, Printers.

PRAELEGENDA.

THE MASSACHUSETTS MEDICAL SOCIETY includes nearly every regularly-educated physician in the State of Massachusetts, and numbers at the present time more than nine hundred members in active practice. For the convenience of the members, the Society is divided into seventeen "District Societies," corresponding to the several Counties, or convenient portions thereof, in the State. The District Societies have frequent meetings for scientific and social purposes; and appoint Delegates or "Councillors" (one for every eight members), who, at stated and other meetings, choose all the officers, and manage all the business of the General Society. Once a year also, in the true Harveian spirit, there is usually "a general feast for all the members; and on the day of such feast, a solemn Oration, by some member previously selected,—to commemorate the dead; to exhort the living; and to advance the cause of science, the honor of the profession, and the fraternal intercourse of the members."*

In obedience to such a call from the Society, the following "Discourse" was prepared. The Author being unexpectedly absent from the country on the day of the "Annual Meeting," the Discourse was most acceptably read by a professional friend. As the Society went without its dinner, in view of the unusual festivities then in preparation for the "American Medical Association" (a National Society), which had accepted an invitation to meet in Boston the following week, the Reader very properly omitted the closing paragraph of the Discourse. But as the paper seems to the Author to terminate rather abruptly without this paragraph, as well as thus to leave the pleasantest portion of the prescribed duties of the day unalluded to, the whole is now printed as it was originally written,—advantage being taken, however, of the opportunity in re-printing, to make a few verbal emendations, which would have been attended to previously had the proof-sheets passed under the Author's supervision in the first instance.

The Society's *caveat* on the next page intimates the risk incurred in deviating from the current course of "opinions or sentiments." Any one, therefore, who feels impelled to do this, must singly take the responsibility; and, if in earnest, be quite ready to reply to objectors, in the words of an Athenian soldier and statesman,—

Πάταξον μὲν ἄκουσον δέ.

Roxbury, Massachusetts,
December, 1865.

---

* The words of Harvey, nearly. See "Life of Harvey," by R. Willis, M.D., London, 1847 [Sydenham Soc.]. See, also, "Oratio Harveiana—a Jacobo A. Wilson, M.D., Londoni, MDCCCL."

At an Adjourned Meeting of the Massachusetts Medical Society, held Oct. 3d, 1860, it was

Resolved, " That the Massachusetts Medical Society hereby declares that it does not consider itself as having endorsed or censured the opinions in former published Annual Discourses, nor will it hold itself responsible for any opinions or sentiments advanced in any future similar discourses."

Resolved, " That the Committee on Publication be directed to print a statement to that effect at the commencement of each Annual Discourse which may hereafter be published."

"These results without doubt will be far from satisfactory ; but of what consequence is that, if they are true ; since, whatever has this character, cannot fail in the end to be of real utility."

Louis, " *On the Effects of Bloodletting*,"
Translated by C. G. Putnam, M.D., Boston, 1836, p. 2.

DISCOURSE.

Mr. President, and Fellows

of the Massachusetts Medical Society:

The profession we follow is capable alike of the divinest endeavor and the meanest purpose. To save it from degradation, and to elevate it to its true position as one of the noblest of human vocations, its faithful votaries have labored with untiring energy in past times and in our own, down to the present hour.[1]*

To understand disease, and to "cure" it, are the great objects and the laudable aspirations of the Medical Profession. The former is difficult; the latter often impossible. Notwithstanding the advanced state of medical science, numbers are at all times prostrate by sickness, and most of the race die prematurely. So uncertain are the effects of dis-

---

* The numerals in the text refer to notes at the end of the Discourse.

cases, and so disastrous often their termination, that
even the simplest attack may become a source of
personal anxiety and alarm. Such, too, are the sym-
pathies of our nature, and so constantly are they
thus called into action, that experienced attendants
upon the sick frequently grasp blindly and fortui-
tously at a multitude of heterogeneous appliances
which have obtained the name of remedies, in the
hope that some one of the number may perchance
rescue or relieve the sufferer. In this way physicians themselves, even the more eminent, are im-
perceptibly and almost inevitably brought to the
practical belief, that, in the officious administration
of drugs sanctioned by custom or prevailing preju-
dice,—the superintending " a course of treatment,"
as it is called,—lies the chief end and aim of their
calling. So thoroughly at last does this idea per-
meate the very life and thought of the daily routine
of our profession, that the mere suggestion of a more
comprehensive and a more scientific view, or a more
rational motive, if suspected of any accompanying
distrust of popular or fashionable professional measures, is liable to be frowned upon as heresy. And
thus it happens, that in every generation the strug-
gle must be renewed to re-establish principles, and to
arrest the mechanical, downward, trade-like tendency of our art ; and good men and true are called
upon, and must be willing, to go to the front and
bear the brunt of the battle.

Grand forward movements in behalf of medical
truth have been made in various directions in our own
time, and with various success ;[2] but it is well known

to the members of this Society, that the great victory of the present century was achieved in this place just thirty years ago. Carefully and irresistibly the first advances were then made, and the first positions gained, until at length the whole argument was driven home, and the stronghold impregnably secured. From henceforth, wherever the English language is spoken or read, the doctrine of self-limitation* will be a ruling influence in the profession until a new era shall require a further advance, or science demand another expression.

Thrown out as a " picket " on this occasion, I will essay my little skirmish and return as speedily as possible to the main column, fortunate if the solitary shot bring down a single resisting error ; more fortunate if it serve to open on any point a clearer view for the progress of the advancing hosts.

" Who did sin, this man, or his parents ?" is a question daily asked, in one form or another, at the bedside of the sick   The frequent response, as well as the query, presupposes, in general, that disease is undoubtedly referable to some indiscretion on the

---

* In the discourse on "Self-limited Diseases," delivered before this Society at its Annual Meeting in May, 1835, by Jacob Bigelow, M.D., the author gives the following definition:

" By a self-limited disease, I would be understood to express one which receives limits from its own nature, and not from foreign influences; one which, after it has obtained foothold in the system, cannot in the present state of our knowledge be eradicated or abridged by art,—but to which there is due a certain succession of processes, to be completed in a certain time, which time and processes may vary with the constitution and condition of the patient, and may tend to death or to recovery, but are not known to be shortened or greatly changed by medical treatment."

part of the sufferer, to the errors of his progeni-
tors, or, at least, to that

> " first disobedience, and the fruit
> Of that forbidden tree, whose mortal taste
> Brought death into the world and all our woe."[3]

To show the fallacy involved in this question, and
that the original answer[4] was the true one,—to show
that disease is not a mere accident in the history of
our race, or due only to unwarrantable experiments
upon our powers of endurance, but rather, that
DISEASE IS A PART OF THE PLAN OF CREATION,—one
of the myriad expressions of Divine thought,—will
form a leading object of the present discourse.

Modern geology has brought to light many wonders of the past. It has revealed to us unmistakable
evidences of the existence on the earth of numerous
classes of organized beings, long ages before the
appearance of the human race.[5] Animals then lived,
flourished, and passed away. Individuals, then as
now, had a limited existence, which death terminated.
Some whole tribes, then as now, were so constituted
that they could live only by the destruction of others.
For this purpose they were provided with organs for
seizing, tearing, and devouring their prey ; while in
some instances they seem to have been armed not
only to destroy but to torture their victims.[6] On the
other hand, organs of defence were furnished to those
in danger of assault, and means of escape given to
the weak.[7] So that it is evident that the same strife
prevailed in those early periods of the world's history as in the present times. In short, there were

voracious mammalia before man, voracious reptiles before mammalia, and voracious fishes before reptiles. Moreover, much curious information has been acquired with regard to the structure and functions of the internal organs of these extinct animals. Not only has the nature of their food been ascertained by the half-digested remains of other animals found within some of these creatures, but the size and structure of the digestive organs themselves, their vascular surface, and the mucous membrane which lined them, have also been made evident by unequivocal marks on the surfaces of their contents.[8] From these and other appearances found in such fossil remains, the inference is unavoidable, that these creatures must have been liable to functional disorders of the abdominal organs similar to those affecting animals of analogous structure at the present day.[9]

While such indications of the nature and habits of these remote animals, and their consequent liability to derangement of function, are thus plainly manifested, the proofs of their liability to organic or structural diseases are complete and unassailable. Extensive enlargements by ossific inflammation have been discovered; as also cavities and outgrowths produced by abcesses. Specimens of caries and necrosis are not infrequent. Other marks of scrofuloid diseases are also recorded. Instances of anchylosis have been noticed; and re-union of fractured bones, with exostosis at the points of junction, have been described and figured. And, more than this, evidences

have been found of recovery from the most extensive
lacerations, involving bony structures, by the fangs
of other animals, where the individual must have
lived long enough afterward to allow the injuries to
be repaired, as far as is ever possible after great loss
of substance.[10]

All these things we have most clearly demonstrated to us in addition to the necessary lethific action
of physical causes, burning, freezing, suffocation,
storms, natural poisons and the like, which also have
existed through all time.

Thus it is evident that, from the beginning (using
here the word in its widest geological meaning, and
not simply in the narrow sense of the beginning of
human existence), life has been subject to dangers,
disorders, and diseases, such as beset it in these latter
days ; and that it has ever had essentially the same
means of escape and modes of recovery. So that we
are led to the inevitable conclusion, that, if the existence and peculiar structure of these ancient animals afford proofs of design, generally acknowledged* to be most wonderful and convincing, so also
their diseases, in the same way made known to us,
and their processes of recovery from disease and

---

* We are aware that the idea of a "Great Artificer" is considered a "fetish-
istic conception," unworthy an educated man or an enlightened age, by some
philosophers, who find an easy solution of all the phenomena of Creation in
"Persistence of Force," spontaneously generated, acting upon matter itself
uncreatable. According to this theory all evils are incidental, to be self-elimi-
nated at some future period. Till a nearer approach of that good time coming,
our manner of dealing with the subject may be permitted, leaving the facts
presented to be translated into other language, should any one ever think it
worth the while.

accident, no art having intervened, must be accepted as equally the result of intelligent contrivance.

In like manner animals now living, whether species continued from former ages, or introduced in more recent periods, all are liable to disease and bodily infirmities.[11] Though they prey upon each other, the numbers thus destroyed probably bear but a limited proportion to those swept away by occasional pestilence. Singly and silently, however, the many, when overtaken by disease, withdraw to some obscure and sheltered nook to await their fate, — of recovery or death. If health returns, they crawl out by degrees to the warmth of day; and many an awkward sportsman has rejoiced over captures due less to his own skill than to the weakness of the convalescing victim. Usually such cases are isolated; and each individual passes away like a falling leaf, unnoticed and unmissed. Occasionally an epidemic rages, and the destruction becomes excessive; while, at times, " diseases of mysterious origin break out in the animal kingdom, and well nigh exterminate the tribes on which they fall."[12]

As it is with wild, so it is with domestic animals. Diseases seize upon them in obedience to laws of which as yet little or nothing is known. Ordinarily they succumb, one by one, unnoticed except by their owners, the scavenger, and the drayman. Now and then, however, the fold is infected, and its future hope endangered.[13] Then the alarm spreads, and the whole country is aroused. In its ignorance and terror it sacrifices life without mercy, and treasure without discretion.[14]

It were well for communities in general to give such subjects more careful study; and especially so for physicians, since "there is every reason," says an eminent authority,[15] "for believing that pathology in man would be greatly benefited by investigations of the diseases of animals."

And so it appears that disease is not only a part of the constant experience of animals, which cannot have any agency in the matter, and only submit to the conditions imposed upon them, but that it obtained in the earliest originations of organized existence, and has continued uninterruptedly to the present time. No "mortal taste," but the will of the Creator, determined and fashioned such a system of diseases, — the evidences of which, foreshadowed in the beginning, become more and more apparent in the subsequent phases of Creation.

Turning now to the human family, whatever may have been its original condition, we find the "lapsed race," from the first pair, brought under the same general scheme. In no period of his life is man exempt from the incursions of disease, from infancy which wakes into an exanthem, to old age which sleeps "sans everything." Every organ has its peculiar diseases, every system of the body its own affections. No forecast or wisdom of the individual can with absolute certainty ward off or delay these attacks. To such an extent is this recognized, that the young adult who has passed through the diseases of childhood, so called, is considered by statists of greater merchantable or insurable value, than one who has still to incur such dangers. Theories have been abundant to show how single diseases may be avoided ; but it does not appear that any disease has as yet been removed from off the globe through man's agency. Flight to the mountains, or to the uttermost parts of the earth, can at no period of life insure perfect exemption, and always at last proves unavailing. We know not even the secondary causes by which diseases are propagated, whether they are atmospheric, miasmatic, or animalcular.[16] They have existed from the beginning, and, so far as we can at present divine, they will continue to exist through all time to come, or until they reach the termination assigned to them. So little are these causes understood, in the usual incursion, spread, and progress of the common diseases of successive years, that not even the wind, that bloweth where it listeth, is less under the guidance or control of human agency or power. Though in all probability obedient to some general law, too subtile to be apprehended as yet, we are utterly unable to predict with certainty what even a day may bring forth of any disease in progress. When an epidemic appears, it often completely confounds all our conceptions of hygienic laws, as well as our preconceived notions of its nature or proper treatment. We cannot tell why it came, or when it will depart ; or whether, under similar circumstances, it will again return. It marches on, often apparently without discrimination, over districts reputed to be healthy ; not unfrequently seizes on purified places, and avoids the polluted; at-

tacks the rich as well as the poor; subverting the theories of the learned and the predictions of the wise.[17]
Now and then we proclaim preventives, destined
only to fail as the announcement escapes our lips.
As we cannot bind the sweet influences of Pleiades,
or loose the bands of Orion, neither can we arrest
the midnight pestilence or the noonday destruction ;
much less can we control in any degree the approach
or progress of those terrific scourges which, in their
appointed times and preördained courses, sweep over
the nations, obeying Him only who rides on the
whirlwind and directs the storm.

Let us take a single case of disease, and observe
what evidences of Design are exhibited in its regu-
lar series of phenomena and modifications. For
example, let us take one of the simplest exanthems.
It is unnecessary to particularize the minuter symptoms. It will be sufficient to notice its general history. For ten or fifteen days, more or less, after
exposure to the morbific cause, called the period of
incubation, the individual, though unaware of his
condition, is as completely under the influence of
the disease as at any subsequent period of its progress. As this stage closes, that of fever sets in,
perhaps with some considerable degree of severity.
Then, in two or three days, an eruption appears, beginning on the face and neck. On the fifth day, it
covers the body and extends to the extremities. On
the sixth it begins to decline on the parts first affected, whilst it is vivid on the general surface. On the
seventh, eighth and ninth, the eruption fades, in the

order in which it came on, leaving the cuticle in a state of exfoliation.[18]

Such is the history of one of the most common exanthems. Others of the class are not unlike it in their general onset, progress, and termination. No evidence of an intelligent contrivance can more distinctly indicate a plan, than that furnished by any of these diseases. In their invasion, incubation, progress, culmination, decline, and disappearance, they are as systematically pre-arranged, and as wonderfully wrought out, as is the life-history of any existence, vegetable or animal, in its conception, embryonic state, infancy, puberty, and adult, middle, and declining age. There cannot be adduced a greater proof of inventive thought, or varied contrivance, perfect in itself, in all its parts and as a whole, than that exhibited in any one of these diseases.

What is thus true with regard to exanthems, is also true of other diseases to a greater extent than might at first be imagined.[19] It may hereafter be shown that a state of incubation belongs to all diseases. One can hardly doubt it in acute cases ; much less in chronic. How seldom is health found to have been perfect up to the very moment of apparent invasion in typhus, and typhoid, and inflammatory affections ! How often rather is it observable that some unappreciated discomfort, or perhaps some exaltation of spirits, is confessed to, when a thorough investigation is attempted ! It is often found that the more obstinate and ordinarily fatal diseases include multifarious weakening disorders, endured by

the victim some time before their true nature is fully
realized. Be this as it may, after a disease has once
taken up its occupancy it follows certain laws peculiar to itself, as the lengthened histories of the books
amply testify. These, though written for an entirely
different purpose, reveal, if we read aright, a remarkable conformity to the idea which we have tried to
develope.

May not what we have shown of the simplest
morbid affections be also true of the more compli-
cated and less understood diseases, such as those of
the blood, for instance, or of the nervous system,
which the acutest observers have failed to explain?
We hear often of metastasis. An internal organ
gives evidence of a severe attack. All at once the
local symptoms abate, and a distant part, an extremity perhaps, becomes the seat of exquisite tender-
ness and intolerable pain. Hardly do these subside
before terrific agony affects the head, and the patient
sinks in the frightful struggles of mania. Theory
explains, that, in the internal organ first attacked,
pus or such-like morbid matter was evolved; that
this, taken up by the adjacent veins, was carried to
the extremity by the veins of that part; and that,
when the last change occurred, it was effected in a
similar way in the direction of the head, — post-
mortem existence of pus in the several parts being
considered proof positive of the truth of the hypo-
thesis. Or else, that the pus taken up by the veins
and carried to the heart, is forced through the
arteries by that organ to the other parts subsequently

affected.[20] But how, in the first instance, can veins whose currents run in the same direction or towards a common centre, carry fluids or other matter in opposite directions? Or, on the second supposition, can we imagine that pus can go unaltered through the whole circuit of the circulation without poisoning the whole system rather than a remote part? Besides, the primitive formation of pus is left wholly unaccounted for. Is it not more rational to think that the original morbific cause, after incubation sufficient to pervade the system, revealed itself, first, in the internal organ, next, in the extremity, and lastly, in the head; the outward demonstrations being only consecutive manifestations of the one unexplained moving cause?

Acute rheumatism may also further illustrate our meaning. This disease, with one central, constitutional morbific cause, shows itself on the outposts in most astonishing ways, — now at the end of one extremity, and in a moment, as it were, leaving that part to appear in a distant one, on the other side of the body. When fixed upon any portion, no one can with any certainty hasten or retard its departure, say how long it will remain, or predict what will be its next point of attack. Each new case is a new enigma. How curiously planned, how varied in uniformity, how singularly wrought out! No finite intelligence could ever have originated such a combination; no human intellect ever approached it in subtilty of contrivance.

But instances need not be multiplied. There is
3

hardly a disease which, if studied in this view, will
not afford an example of wondrous designing power.
All the resources of art would be unavailing in an
attempt to originate even one of the simplest specific
diseases. Great is the mystery that overhangs the
nature of morbific causes. The highest intellects
have proved incompetent to its solution. Volumes
have been written to elucidate it, still the mystery
remains as obscure as in the days of the earliest
observers. But the inference is unavoidable, that,
if the human mind cannot unfold the marvels of a
disease with all its attendant and antecedent phenomena, and much less find its morbific cause, it must
have required a higher intellect, than any created, to
have combined these agents, and arranged the laws
by which they all are governed.

Not less worthy of notice are the different susceptibilities of different individuals to any single disease ;
and of a single individual to different diseases. When
the seeds of disease are scattered abroad, many fall
into unprepared systems, and after springing up,
quickly wither away ; not every acorn becomes an
oak. Let a large number of persons be simultaneously exposed to contagion : one portion of them
would soon sink under its influence ; another would
be severely affected ; still another, and perhaps the
largest, would suffer moderately ; while a few
would pass unscathed, entirely unaffected by its presence. In a great number this susceptibility would
be exhausted by one attack, so that the subjects of it
could bear any amount of subsequent exposure with

impunity. On the other hand, a few would receive the disease a second, and some even a third time. This difference of susceptibility obtains in regard to most if not all diseases, and to the protective power in many, — how many, is not yet fully ascertained. In other cases, however, one attack only predisposes to a repetition. In this respect, also, there seems to be a graduated scale, arranged with forethought and planned by intelligence. And all this is true no less of individuals than of classes.

Again, diseases are distributed through the different seasons of the year with such a degree of constancy, that the seasons themselves are sometimes spoken of as the causes of the diseases. But a little reflection will enable one to see, that, in the nature of things, there is no essential or known reason why diseases of the bronchial mucous membrane should prevail in winter, or those of the intestinal mucous membrane in the summer ; why the plague should prefer heat, and variola cold, for its devastations.

In like manner the appearance and peculiar characteristics of common diseases in ordinary seasons, or the severer cases of epidemics in all seasons, may never be satisfactorily accounted for by the external surroundings of the victims. After most careful investigations, writers are still compelled to admit that there must be some unknown condition, some cause not understood, other than the poverty, privations, filth, and locality of those attacked.[21] The simple explanation is to be found in the idea of an original Plan, as we are attempting

to demonstrate. With this as the guiding idea, how
much more intelligible become such investigations
of disease ; how much easier the unravelling of the
laws which govern organized existence ; how much
time saved, now lost in a fruitless search for specific
causes !

Other evidences of Design and fixed law may be
noticed in the general averages of sickness and
mortality. These are such and so constant, that in-
surers can calculate. with accuracy their probable
losses from one decade to another, though their
patrons are selected from the most vigorous and
favored classes ; and any community can estimate,
if it chooses, its loss of time by sickness, so as to
provide beforehand for the coming emergencies of
future years.[22]

The geographical distribution of the various animals and plants within certain limits,[23] a discovery
which has given additional interest to natural science in our day, is not more remarkable than the
geographical distribution of diseases.[24] While some
of them seem to be almost cosmopolitan in their extent, others are confined to restricted localities, beyond
which, without any apparent reason, they seem un-
willing to go. As some plants thrive best in connection with others, or in near proximity, so there are
diseases which seem to have a mutual affinity, or
appear generally in connection with each other ;
while, on the other hand, some unexplained antago-
nisms' and complements exist among diseases, not
unlike those observed in the vegetable kingdom.[25]

Furthermore, that power peculiar to organized beings, which enables them to endure, within wide limits, all kinds of physical changes and exhausting influences, is no less remarkable in the tolerance of diseases. This " reserved force " seems a necessity preliminary to the possibility of disease, or at least to recovery from it. Without this reserved force, ordinary functions would be in constant danger of interruption or absolute destruction. With it, the severest malady may pass through all its stages to perfect recovery, without, in the end, greatly injuring the individual. This will appear a more remarkable provision when we consider, that during disease the ordinary supply of nourishment for the development of force is refused by the patient, and often only so much is accepted as may be barely sufficient to continue existence. We do wrong to call this, or any phase of it, a *vis medicatrix*,— a term (the sooner discarded the better) involving theories long since abandoned, and now almost forgotten. It is simply a vital principle of endurance, sustaining the organism through all the period of disease, as necessary at the outset as at the close.

Such are some of the evidences of forethought and design in the introduction of diseases. These evidences are to be seen in the fossil remains of animals which lived and suffered long before man appeared upon the earth. They are to be seen in the diseases of animals now existing,— in the wild which avoid, and in the domestic which cling to, the

abodes of the human race. They are to be seen
more universally, and more completely, developed
in man himself, as, from the cradle to the grave, he
passes through one experience to another by allotted
stages. They are to be seen in the histories of
separate diseases so systematically and mysteriously
constructed; and in their geographical distribution,
periodicity, and modes of onset and of decline. Additional evidences are also to be seen in the different
susceptibilities of individuals, and in the power of
endurance possessed by all. From whatever point
the subject is viewed, multiplied evidences arise of
intelligent and inventive authorship. On all sides
*these evidences are of the same kind as are adduced
to show Design in other operations of nature;* and,
if admitted anywhere, we must admit its manifest
revelation in the devising and the orderly contrivance
of diseases.*

Thus it appears that the idea of Diseases must
have originated in the Creator's mind, and its development formed a part of the Plan of Creation from
the beginning. The ultimate purpose of such a
Plan it is not for man to determine.[26] Deliberately
devised, diseases do not necessarily imply "gratuitous

---

* N. B. The limits of this Discourse admitted only of a restricted development of the argument from diseases in animals; that from diseases in the
vegetable kingdom, exhibited in every forest, grove, and orchard, is equally
impressive and convincing.

For the same reason idiopathic diseases only (those "realities" manifested
in "a series of consecutive changes") have been considered. Disorders
(irregular or disturbed performance of function) afford equally good illus-
trations of plan, in the laws which govern them, and in the subsequent
restoration from their effects.

malevolence"; for, despite of some philosophers, it
is quite possible to conceive of the earth and all
that is therein, simply as an expression of Divine
thought, without reference to the question of good
and evil. But to repel such censure is foreign to
our present purpose; *whatever is*, is enough for us
as scientific men now to consider, humbly acknow-
ledging, that " in the Divine government the matter
of fact always determines the question of right, and
that whatever has been done by Him, who rendereth
no account to man of his matters, He had in all ages,
and in all places, an unchallengeable right to do."[27]

Such being the facts, though it may never be
explained why organized existences always have
been, and until a new order of things has arisen
always will be, subject to diseases, yet the extrication
of what cannot be explained from what may be, is
no small addition to any science. Recognizing such
limitations, we shall not, like the great men whom
Hippocrates so ingeniously refuted (for there were
great physicians before his time), labor to refer all
the afflictions of the race "to hot, or cold, or wet, or
dry;"[28] nor to "figments called inflammations," which
have been so quick to disappear under the tests of
our own day; nor to any of the many other theories which between those have had their short-lived
career. But we shall consider the causes of diseases
to be primitive purposes, as much so as electricity or
gravitation, and, proceeding as we do with those
subjects, we shall study their development, and the

phenomena to which they give rise, with more satis-
faction to ourselves and benefit to the sick, and with
the positive enlargement of our science.

Since these things are so, it may perhaps be said
that we may as well fold our hands, and resign ourselves with indifference to whatever fate may befall
us. By no means. The storm may arise and the
winds may blow, but we may seek shelter from the
former, or wrap our mantles closer to exclude the
latter. Even against the inconveniences of a summer
shower we may oppose the delicate contrivances of
modern invention. But it does follow that we may
not attempt to attack the laws of nature with any
hope of arresting the fury of the elements, or the
influence of their disturbances. We may indeed estimate their forces, calculate their movements, and,
having possessed ourselves of all that is known of
them, govern our conduct so as in many cases to avoid
them in the outset, or at least to mitigate the evils
in their train, or to take advantage of whatever of
good can be derived from their presence.

So with regard to diseases, we may not have it in
our power to banish their elements from existence;
we may not often prevent their coming, or be able
to stay their progress; we may not jugulate or break
them up at pleasure when once they have seized
upon us, or greatly shorten their continuance; we
may not amend their destructive characters, or very
sensibly diminish average mortality;[29] — nevertheless, suitably recognizing their place in the great
Plan of Creation, and acquiring as full a knowledge

as possible of their phenomena, with a just estimate of human power, we may seek, with some certainty of success, to evade their approach, or to save our- selves from many of the inconveniences and dangers of their attacks.[30] Thus, when a disease "has ob- tained foothold in the system," we may remove as far as possible obstacles to the natural progress of its "succession of processes," and sustain the system as well as may be in its power of endurance, until these processes are duly and safely completed. To do this to perfection, is no easy matter. It will re- quire a greater knowledge of disease than any indi- vidual, however learned, has yet acquired; a more thorough investigation of each separate case than is now made by the most pains-taking practitioner; a more complete mastery and discriminating use of all the appliances of our art than has been heretofore possessed and practised; and a more absolute and abiding control of the patient and his surroundings than was ever yet granted to any medical attendant.[31] Possibly the Profession and the Public may hereafter be educated up to such a state of perfection in the management of the sick; though, as yet, even the profession seems not quite willing to fully accept all that is now known of the nature and laws of disease.[32]

"The physician," says the learned translator of Hippocrates,[33] "who cannot inform his patient what would be the probable issue of his complaint, if allowed to follow its natural course, is not qualified to prescribe any rational plan for its cure." But how small a proportion of the profession could con-

sistently practise their calling for a single day, were
this test strictly exacted! Who among us ever saw
a disease allowed to follow its natural course to its
termination, unless, bolder than his neighbors, he
risked the denunciation of his peers and dared by
himself to try the experiment? Yet the trial is not
so dangerous as was formerly believed;[34] and, if
entered upon as unhesitatingly and with the same
confident expectation with which newly vaunted
remedies are often given, more "wonderful cures"
would be witnessed than were ever related in the
books. The time is coming, perhaps it is nearer
than we are aware of, when the public shall no
longer consider the proper care of the sick (their
true *cure*) to consist in a mysterious and indispensable administration of drugs, but in rationally and
understandingly attending to all their necessities;[35]
when the young aspirant for patronage shall not find
it necessary, in order to satisfy the bystanders, to
write his recipe before he has examined his patient,[36]
or to authoritatively announce the name of the disease before he has had time to comprehend the
symptoms; — and there is no reason why the profession should not now, by lofty endeavor and combined action, strive with success to bring about such
a desirable result. When this is accomplished, the
not unreasonable requirement above quoted may be
fully accepted. At any rate, it is time that the education of pupils in the study of disease should be
founded on a new basis. Not a school in Christen-
dom ever yet afforded proper opportunities, if any

at all, for studying the natural course of diseases.[37] Under different teachers, if we may credit eminent authorities and our own observations, the same disease may assume different outward appearances, according as the several " courses of treatment " may differ from each other.[38] Although the immediate effects of drugs, and their strictly therapeutic influence, if any, are very different matters, and ought never to be confounded,[39] — students are too often led to believe that all the recoveries they have seen have been due to the prescriptions selected ; and they go out into the world under the apprehension, that, if they do not generally " cure " disease, it will be from not having the good fortune to hit upon the right course of medication. The exhibition of a multifarious mixture, in order perchance to include the right ingredient, is not merely a fitting, but the most obvious, corollary to their previous instruction. To most men, years of anxious and much unsatisfactory experience ; to some, a whole life of disappointment ending in utter scepticism of the value of medicine, are the results of such erroneous beginnings.[40]

The doctrine we have advanced and advocated leads to a different procedure. It leads to an abandonment of the old notions of the primary causes of diseases. It leads to a new view of the purpose of diseases themselves. It shows the idea untenable, that disease, an evil, is to be expelled from the system by some antagonistic power only, the *vis medicatrix*, for example ; or by a new and incompatible disease artificially induced ; or that it is in itself an effort *(conamen)* to expel from the body an enemy already in possession;[41]—but that it is one of the attendants of life, instituted in the Beginning. And, ignoring none of the real acquisitions of the past, this doctrine divests the truth of many of the errors which have thus far impeded its progress.

This doctrine being accepted, the proper acquisition of our art will demand of students, in the first place, a thorough knowledge of the body in its healthy condition,—its organic structure, its outward form,[42] and its internal functions ; and, secondly, the investigation of the natural phenomena of disease *undisturbed by medication*, as a necessary preliminary to its proper management. It will require of them also a careful study of the operations of the mind as affecting the body, and their mutual reactions upon each other, in health as well as in disease, — health and disease being parts of one great Plan, and often intricately involved in each other. In these directions medical education has been deficient, and subsequent attention in after-life remiss. Let coming students take warning from the deficiencies and failings of those who have preceded them.[43] Let them, thus properly grounded, and not till then, proceed to study all the effects of accepted remedial agents. Every step from such a base will be a true progress for themselves and their science, no disappointment or scepticism ensuing. Every advance in this way will be in the right direction,— *vestigia nulla retrorsum.*

Of late years it has been quite common to vaunt

the power man may have over plants and animals
in modifying their form, color, growth, and other
qualities, and to adduce this as an argument in favor
of a similar power over diseases. But the two
cases are far from being analogous. It is one thing
to raise a few deformed sheep, or to increase the
number of vertebræ of birds from generation to
generation,[44] by " selective breeding," so called ; and
quite another thing to modify the course and termination of disease in a particular individual. They
are separate matters, connected by no logical se-
quence. The one is necessarily limited to the life of
an individual, or only to the duration of a disease in
an individual ; the other may, nay, must extend
through successive generations, or successive ages.
Besides, the permanency of species has not yet been
disproved, and it will be time enough to use such
arguments when dogs shall be actually bred from
wolves, or an ape be unquestionably transformed into
a human being.

While admiring the activity in our medical schools,
the facilities of instruction and for clinical observations at our hospitals, the zeal of societies, the energy
of individuals, and all the various helps to professional advancement, now so multiplied and abundant, one
cannot but regret the still prevalent tendency to recur
so readily to second causes, and to impede the advancement of medical science by claiming for it
more than is consistent with actual truth. The
medical press, so often boasted of as the great dis-
seminator of medical knowledge, is still too often

the vehicle of false philosophy and unworthy assumptions. False facts, false reasoning, and non-
sequitur conclusions fill up a large portion of peri-
odical publications. Even the more stately volume
seems incomplete without its remarkable cases selected for an object, and its infallible formulæ, which
perhaps have never had a trial.[45] An author who
shall candidly relate his own experience, in ordinary
cases, of expectations disappointed and unsuccessful ·
issues following the employment of reputed infallible agents (and such experience only), will richly
deserve, if he does not receive, the thanks of the
profession, and be indeed " more than armies to the
common weal."[46]

Brethren :

Fashions in medication are fluctuating and fleet-
ing. Each age flatters itself that it has made a
great advance over the previous one, and has reached
at last something established and permanent. But
we smile at the notions of our predecessors, only to
be laughed at by those who come after us. Time
was (men are living who remember it) when pneu-
monia was considered a fatal complaint, unless sub-
dued by venesection at its onset ; now it is instanced,
by an eminent observer, as the purest example of a
self-limited disease.[47] Time was (physicians are with
us who thus practised) when spasmodic croup, so
called, was believed to be an imminently dangerous
disease if the external jugular vein were not imme-

diately opened; now it is known to be a compara-
tively harmless accompaniment of another disorder,
and needing in itself no special interference. Time
was (our own day embraces it) when it was publicly
taught, that mercury given to salivation was not only
the specific, or antidote, for iritis, but absolutely essential to its successful treatment; now, one of our
number has been justly called a public benefactor
for showing that such practice is not only unnecessary, but often grievously detrimental in that affec-
tion.[48] But why multiply examples? So it has been,
and so it will ever continue to be, until more correct
views are acquired of the Plan of Creation, and of
human powers under it. The great facts of our
science are permanent, and, however feebly stated
from time to time, or hesitatingly received, will at
last prevail and triumph. False assumptions are
dangerous expedients, and the most ignorant will
ever be the most likely to practise upon them.[49]
Truth is weakened by any addition of error; and the
profession that allows it must in the end abandon its
own self-respect. The remedy is in our own hands;
let us be heroic enough to apply it in season.

" Medicine," says our American Hippocrates, " is
the art of understanding diseases, and of curing or
relieving them when possible."[50] To this sage remark
it may be added, that a Doctor of medicine should
also teach the patient and his friends to acquiesce
in an intelligent submission to the laws of disease;[51]
laws as manifest and inflexible as those of health.
This done, the Profession will acquire a dignity be-

fore unknown to it; and the Attendant will become
an enlightened guide, instead of an uncertain and
bewildering dealer-out of nostrums.

To turn increasing attention in the direction indi-
cated, we ventured on the perilous duty of to-day.
Let us hope, that, as impediments are one by one
removed, progress may be easier in time to come.
There is nothing in time past to discourage renewed
effort. Though yet afar off, the goal is nevertheless
in sight. The present time is propitious. Allied
sciences are on the move. It is for us to hasten on,
and to display our standard in the foremost ranks.
Thus shall we better satisfy the demands of the age,
and truly ennoble our Profession.

BRETHREN:

During the past year, twenty-two of our number have yielded to the common fate of mortality.
The Secretary's list, which he read to-day, has
given us, name by name, the melancholy announcement. The courteous Bartlett, the munificent
Walker, the genial and true-hearted Coale, the brave
and tender-hearted Sargent, and our other martyrs,
Fox, Heath and Hoyt, with other well-known and
cherished friends, have gone to their rest. We strew
the fresh-formed mound with cypress mingled with
laurel, and kindly drop the tear of friendship, as,
imitating their example, we press earnestly forward
to the struggles awaiting us. Faithfully and loyally

they served the cause of humanity and of their country; ardently would we recount their virtues, and for-ever hold them in honored remembrance.

And now, Brethren, having attempted the weightier duty of the hour, and paid a tribute to the memory of those who have "passed over to the majority,"[52] let us, as the quick-step follows the dirge and the volley, hasten with cheerful pace to the pleasanter task reserved for us at the festive board. In all ages weariness and sorrow have in this way sought appropriate relief. Thus it is that Old Homer closes the solemn meeting of Priam and Achilles; and fortifies the latter with a stronger instance in justification of his own course. Let us, in obedience to a common nature, imitate such high examples, and becomingly conform to the time-honored custom of our Society, —

> Now feast we, — not in grief's severest mood
> The bright-haired Niobe forgot her food;
> Then let us now, ere coming cares annoy,
> Our thoughts awhile on needful food employ;
> To-morrow range the vales and hill-sides o'er,
> The weak to succor and the dead deplore.[53]

5

NOTES.

1. Page 5, Line 7.

"Il faut toujours en revenir à cette triste vérité, que la médecine est la plus noble des professions et le plus triste des métiers."—*Gaz. Med. de Paris*, 1851, Tom. v. p. 418.

"Medicus sum, non vero formularum medicarum præscriptor ; quas ego duas, sive artes, sive dotes, sive etiam provincias, appellare libeat, toto cœlo a se invicem distare arbitror." Sydenham, "*Diss. Epist.*," §42.

Said an esteemed friend, as we left the hall at the last annual meeting, "So, you read next year ; well, don't give us any of your heresy!" The exordium of this Discourse was written that evening.

2. Page 6, Line 32.

See a list of publications in "Expositions of Rational Medicine," by Jacob Bigelow, M.D., Boston, 1858, pp. 57-60. Several other well-known papers have been published since that date.

3. Page 8, Line 5.

"Upon his taking his place as lord of the terrestrial creations, a specific injunction was given, guarded by a penalty for its violation : 'In the day thou eatest thereof thou shalt surely die.' But, having taken upon himself the fearful responsibility of casting off the authority of his rightful sovereign, he came to disregard all wholesome laws, whether outspoken from the cloud upon Sinai, or written upon the organism of his physical nature ; hence the insane perversions in physiology and psychology, including the poisoning of the senses of taste and smell, those faithful guardians of life and health and beauty ; and hence the thousand forms of disease that flesh is now heir to."—"*Health : its Friends and its Foes*," by R. D. Mussey, M.D., LL.D., Boston, 1862, pp. 190, 191.

"Hence the disorder and disease ; hence the groaning and travailing together of the whole creation ; it is all the supernatural work, the bad miracle of sin." *Bushnell's "Nature and the Supernatural,"* New York, 1861, p. 218.

"Medicines are created by our offended God to relieve diseases which all originate in sin." "*Scott's Commentary, Matt.—John,*" p. 650, Philadelphia, 1860.

Miss Nightingale, whose influence exceeds in effect a score of such writers, takes a diametrically opposite view of these matters. She speaks of diseases as "conditions, like a dirty and a clean condition, and just as much under our control, * * * conditions in which we have placed ourselves;" and seems to think that we can originate diseases at will. She says:

"I have seen with my eyes and smelt with my nose smallpox growing up in first specimens, either in close rooms, or in over-crowded wards, where it could not by any possibility have been 'caught,' but must have begun."

"Nay more, I have seen diseases begin, grow up, and pass into one another." "*Notes on Nursing,*" p. 26, *note.*

Disrespectful as it may seem, one can hardly repress the exclamation, "Oh, Gammer, what big eyes you've got!"

4. PAGE 8, LINE 7.

We quote the answer, for the benefit of those who may not remember it! "Jesus answered, Neither hath this man sinned, nor his parents; but that the works of God should be made manifest in him." JOHN ix. 3.

5. PAGE 8, LINE 18.

See Hugh Miller, "Testimony of the Rocks;" Buckland, "Reliquiæ Diluvianæ;" "Bridgewater Treatise;" Mantell; and others.

6. PAGE 8, LINE 26.

"Fossil sharks, with weapons so murderous, that they must have been, according to Agassiz, the pirates of that period." MILLER, "*Old Red Sandstone,*" p. 215.

7. PAGE 8, LINE 29.

For a description of the sting of the Pleuracanthus, offensive organs, and defensive armor of other animals, see Miller, "Testimony of the Rocks," pp. 99 *et seq.*

8. PAGE 9, LINE 12.

See Buckland, "Bridgewater Treatise," pp. 187-201.

9. Page 9, Line 18.

Buckland, *ib.* p. 190, *note.* " The quantity of this coprolite is prodigious, when compared with the size of the animal in which it occurs ; and, if we were not acquainted with the powers of the digestive organs of reptiles and fishes, and their capacity of gorging the larger animals that form their prey, the great space within these fossil skeletons of Ichthyosauri, which is occasionally filled with coprolitic matter, would appear inexplicable."

10. Page 10, Line 6.

Mr. Clift's case, see Buckland, " Reliquiæ Diluvianæ." p. 71. Cuvier, " Ossemens Fossiles," Vol. iv. p. 396, and plate. See also Zies, " Beschreibung mehrerer kranker Knochen vorweltliche Thiere," Leipzig, 1856. A résumé, with additional descriptions of specimens in the Dresden Collections.

11. Page 11, Line 6.

According to Dr. Livingston, "many diseases prevail among wild animals " in South Africa. "*Researches,*" p. 149.
See " Recherches de Pathologie Comparée," Ch. F. Heusinger. Cassel, 1818.

12. Page 11, Line 22.

Hugh Miller, " Old Red Sandstone," p. 222; which see also for several noted instances of epidemics.
See "Traité d'Hygiène Agricole, par F. A. Rufener." 8vo., Fribourg, 1858.
See also a valuable work, " Die Einimpfung der Lungenseuche des Rindviches," &c. By Prof. J. M. Kreutzer, 8vo., Erlangen, 1854.

13. Page 11, Line 29.

" Spemque, gregemque simul, cunctamque ab origine gentem."
Georgic, iii., l. 473.

14. Page 11, Line 32.

" At length she strikes a universal blow ;
To death at once whole herds of cattle go."
Dryden's Georgics, iii., lines 827–8.

15. Page 12, Line 4.

Sir James Forbes, M.D., " Nature and Art in the Cure of Disease," p. 46.

16. Page 13, Line 11.

Sir Henry Holland, M.D., "Medical Notes and Reflections," Chap. xxvi. T. Thompson, "Annals of Influenza," p. 385; and others.

17. Page 14, Line 2.

In November, 1818, cholera broke out, nearly simultaneously, in two vessels in mid-ocean, about a thousand miles apart, one sixteen days out, and the other twenty-seven, from an unaffected port. "*British and Foreign Medico-Chirurgical Review,*" No. lxxii., pp. 414–5.

In the summer there was a severe outbreak in the island of The Grand Canary. No other of the group was affected. The origin of the disease could not be traced.—*Ib.* p. 417.

In 1832, a vessel from New York to Newport, carrying a cargo of disinfectants, and being completely saturated with their odors, had her crew attacked with cholera, at sea, and lost several on the passage, or on arrival. This we have on indubitable authority.

"It is a fact that the Asiatic cholera twice spared the poor Jews, in 'The Ghetto,' who live most crowded, filthily, and with bad nourishment." *Letter to the Author from* Dr. Valerj, *of Rome, Italy.*

Like instances abound in all authors on such subjects.

18. Page 15, Line 2.

Marshall Hall, "Theory and Practice." Article, *Measles.*

19. Page 15, Line 20.

"The poison which generates cholera" "certainly possesses in an extraordinary degree the properties, which all other morbid poisons possess in some degree, of lying latent for a length of time, in certain localities, or in the constitutions of individuals, or both," &c. "*Cyclop. Pract. Med.,*" Vol. iii. p. 253.

20. Page 17, Line 1.

This case occurred while writing this part of the Discourse. The explanations given are those of the eminent gentlemen in attendance. Copland says of another case of Metastasis, "The transfer was instantaneous, * * * the medium being evidently the nervous system." "*Dict. Prac. Med.,*" Article *Disease,* §173A.

21. Page 19, Line 30.

"Cyclopedia of Practical Medicine," Vol. iii. pp. 23 *b,* and 251, *et seq.*

22. Page 20, Line 16.

So stated to me by agents of known ability. See also Reports to the Legislature on Insurance, &c. Memorial of the Boston Sanitary Association, pp. 9 et seq., Boston, 1861.

23. Page 20, Line 18.

First sketched in its great outlines by Humboldt, and most fully demonstrated for the class of mollusks in their distribution along our coast by our President, Dr. A. A. Gould, in 1840. See " Invertebrata of Massachusetts," p. 315. Also " Proceedings of Boston Nat. Hist. Soc.," Vol. iii. p. 483. " U. S. Exploring Expedition, Mollusca," pp. 9 et seq.

24. Page 20, Line 21.

" On peut donc dire avec une parfaite exactitude, des maladies, considérées au point de vue géographique, comme des végétaux, qu'elles ont leurs habitats, leurs stations, leurs limites, sous le triple rapport de la latitude, de l'altitude et même de la longitude géographique."—Boudin, " Traité de Géographie et de Statistique Médicale," Paris, 1857, Vol. ii. p. 227.

25. Page 20, Line 31.

Cretinism denotes Goïtre in the same country. In Central Europe, typhoid fever accompanies phthisis.—Boudin.

Intermittent fever and phthisis are not usually prevalent in the same locality.

Wherever Calopogon is met with, one may expect to find Arethusa in close proximity.

The thistle is destructive to oats ; erigeron, to wheat ; scabious, to flax.

In the United States, some diseases (phthisis, for example) diminish from the North to the South, while others (abdominal fevers) increase in the same direction.— Dr. A. A. Gould, Résumé U. S. Census, 1860, in Massachusetts Registration Report, 1861, p. 53, and 1862, p. 48.

26. Page 22, Line 23.

Dr. Brown thus quaintly states a popular belief : " A brisk fever clarifies the entire man ; * * * it is like cleaning a chimney by setting it on fire ; it is perilous, but thorough."—" Horæ Subsecivæ," p. 100 ; " Spare Hours," p. 206.

Said El Hadgi the Fakir, quite as sensibly, " Welcome the disease, if it bring thee acquainted with a wise physician. For saith the poet, ' It is well to have fallen to the earth, if, while grovelling there, thou shalt discover a diamond.' " — " Chronicles of the Canongate," Vol. ii. p. 139.

27. Page 23, Line 12.

Hugh Miller, "Testimony of the Rocks," p. 104.

28. Page 23, Line 23.

Hippocrates, "Ancient Medicine," §15, Ed. Sydenham Soc. "Argument" by Dr. Adams, *ib.* p. 158.

29. Page 21, Line 29.

" La proportion des décès est loin d'avoir diminué avec l'accroissement du nombre des médecins."—Boudin, Vol. ii. p. 81.

30. Page 25, Line 5.

"Dit M. Quetelet, ' L'art de guérir exerce peu d'influence sur le nombre des décès, mais il en a beaucoup pour améliorer physiquement le peuple. Il diminue la somme des douleurs,' &c."—*Ib.* p. 86.

31. Page 25, Line 20.

Any one who may fear that his occupation will be gone, should he admit the possibility of treating disease without drugs, will find the daily routine of a Rational Physician well set forth in the following extracts:
" The medical man will find ample scope for the exercise of his faculties, even in cases where special drugging may not be requisite. Close attention, acute observation, and the expenditure of not a little time, will be indispensable on his part, in order to effectually act upon modified health,—laws in regard to rest, the many nice points connected with diet, the hygrometric condition, temperature, and free circulation of the air, change of air, clothing, cleanliness, &c. His attention must also be directed to exciting or aggravating causes of disease in the locality, the residence, the room, or the person of the patient. He will, moreover, have to take care that the mind of the sufferer is kept in as tranquil a condition as circumstances will admit of," &c.—"*Rational Medicine ; The Hunterian Oration for 1860,*" by S. H. Ward, M.D., &c., p. 48.
" What we desire is a statement of the excess of benefit derived . . . . over that from careful treatment without drugs (which we pray for, should we have cholera ourselves) and over other systems of medication. When we say careful treatment, we mean the giving of nourishment, and the comforting the patient in those small details which can only be attended to by a kind and skilful Medical attendant, who does not conceive himself bound to *try* something for the sake of appearances." *London* " *Medical Times and Gazette,*" Nov., 1865, p. 577.

The idea that "according to the author's theory of disease," and we may add or that of any other, except, perhaps, the Russians' (see note 37), "judicious medical treatment," at the hands of an enlightened and "careful" physician, "is of no use," has been justly characterized by a writer in the Boston Medical and Surgical Journal, Nov. 23,'1865, p. 336, as being "*as absurd as it is untrue.*"

32. PAGE 25, LINE 25.

"If what is really *known* of the laws of disease were told to the members of the profession, more than half of them would indignantly discredit it," said an eminent pathologist to the author a few months since.

33. PAGE 25, LINE 27.

Dr. Francis Adams, LL.D., "Life of Hippocrates," p. 18. *Sydenham Society's Edition.*

34. PAGE 26, LINE 7.

For "baneful effects" of trusting to nature, see Cullen's preface to his "Practice of Physic."

35. PAGE 26, LINE 17.

"He who gives the least medicine, and that of the least offensive kind, is coming to be regarded as the best physician. It is, by the intelligent head of the family, held no impeachment of a physician's skill that he leaves no recipe, and directs measures so simple as to reflect no mystery on his craft."—*Boston Post* (*newspaper*), July, 1864.

36. PAGE 26, LINE 20.

Baglivi, Hippocrates *Romanus* ab Aliberto vocatus, ait: "In curatione morborum, qui morum aliquam admittunt, hoc ordine progredior. Primâ die totus sum in examinando, &c. * * * Secundâ die, diligentius consideratis rebus antedictis, morbi speciem tandem decerno, et exinde remedia opportuna præscribere incipio."—*Prax. Med.*, p. 110.

37. PAGE 27, LINE 1.

Possibly there may be an exception in Russia. "Dr. Hawrowitz, Physician to Prince Constantine, told me," said Dr. Rocser, Physician to the late King Otho, to the author, "that the mortality in the hospital of the old Russians at Moskowa,—who consider by their faith disease as a punishment by God, and the application of medicines for that reason a sin,—

is not greater, if not less, than in other hospitals. They apply only cleanliness and good nourishment."—*MS. Notes of a Visit to Athens, Greece*
1860.

The result thus stated agrees with the author's experience. In the epidemics of 1847-8, he took care of over three hundred cases of typhus
fever without administering drugs. The cases were taken indiscrimi-
nately, including those in a dying state when first seen. The result was
thirty-one deaths in three hundred and seven cases.* In an epidemic
of scarlet fever in 1848-9, out of eighty-one cases so cared for, seventy-
seven recovered. With every attention to the comfort of the sick and as
thorough nursing as possible, the progress of the disease was as tolerable
as, its continuance as short as, and dangerous sequelæ less frequent than
in other cases more "actively treated." In 1849, of forty cases of measles, thirty-nine recovered. The author sometimes takes care of the more
painful diseases, rheumatism for instance, without drugs. It requires
greater patience and painstaking on the part of the practitioner, but the
result is satisfactory. "I had not time," said a prominent physician the
other day in the author's hearing, "to persuade the family that the patient did not need any medicine, so I wrote a prescription and departed."

38. PAGE 27, LINE 6.

" Nam sæpe accidit ut facies morbi variet pro vario medicandi processu,
ac nonnulla symptomata non tam morbo, quam medico, debentur."
SYDENHAM, "*Observationes Medicæ,*" §10.

Baglivi and others have similar expressions.

"The constant interference of art, in the form of medical treatment,
with the normal processes of disease, has not only had the frequent effect
of distorting them in reality, but, even when it failed to do so, has created the belief that *it did so;* leading in either case to an inference equally
wrong — the false picture, in the one instance, being supposed to be true;
the true picture, in the other, being supposed to be false."— Sir J.
FORBES, "*Nature and Art,*" p. 6.

39. PAGE 27, LINE 9.

" Objections may still be made to the inferences, which I think may be
rigorously deduced, from the fact that patients attacked with erysipelas
of the face are very often sensibly relieved, have much less redness of

---

* "The proportion of deaths from typhus may seem large (427 in 2009 cases), yet so fatal
is that disease, that, on comparison with the medical statistics of such other hospitals as the
Board have at present the opportunity of examining, the practice at Ward's Island has been
among the most successful." *Annual Reports of Commissioners of Emigration, State
of New York,*" 8 vo. pp. 100-101. (*Report for* 1851.)

face, during, or immediately after the bloodletting than before. This relief and paleness of the face do indeed take place sometimes; but these effects are momentary, and the progress of cure is not more rapid in these cases than in others. So that the only conclusion from this fact is, that the immediate and the strictly therapeutic effects of remedies must not be confounded."—Louis, "*On the Effects of Bloodletting,*" &c., Dr. Putnam's Translation, p. 17.

"In about a thousand cases [of Plague] * * * although the medicines produced their wonted effects upon the organism, the malady neither ceased nor changed." "*Cyclopedia of Practical Medicine,*" Vol. iii. p. 552, b.

40. PAGE 27, LINE 22.

Or, possibly, some abate at last their *hyperpraxis,* and adopt in part a more rational method ; and, finding that diseases pass off, to say the least, quite as readily as under the previous management, and with fewer severe or abnormal symptoms, they comfort themselves with the absurd conclusion that disease has changed its character during their short day and generation.

"The failure of the various medications advocated," said Velpeau in reply to Le Verrier, at a recent meeting of the Academy of Sciences, "is attributable to a grievous mistake of the public, and even of professional men, who are under the impression that diseases are not susceptible of a spontaneous cure. Many affections yield without treatment, and, it must be acknowledged, sometimes in spite of all treatment. To this fact we must not be wilfully blind. An opposite opinion unfortunately prevails . . . . . After the exhibition of a remedy the symptoms have yielded once, twice, thrice, or oftener ; hence it is inferred that the cure has been the consequence of the treatment. The inference is a natural one, but almost invariably incorrect."—"*Journal de Médecine et de Chirurgie Pratiques,*" Nov., 1856, art. 6981—Eng. Ed.

41. PAGE 28, LINE 2.

" 'Vis medicatrix naturæ' is a favorite professional expression, a time-hallowed portion of medical phraseology. * * * Is there indeed, among other wonders of our corporeal being, a subtile force, inherent in the very organization itself, whose office it is to protect vitality, in its very arcana—to correct errors of function, and restore lesions of structure? So our accepted phraseology implies."—"*Address before the Kentucky State Medical Society,*" by J. B. FLINT, M.D., Pres't Soc., 1859, pp. 8-9.

" The conversion of the original disease into another is occasionally salutary. * * * It is a very common object of art to produce this kind of conversion."—"*Pract. Principles of Medicine*," by J. Conolly —"*Cyclop. Pract. Med.*," Vol. iii. p. 272.

"Dictat Ratio (si quid ego hic judico), Morbum, quantumlibet ejus causæ humano corpori adversentur, nihil esse aliud quam Naturæ conamen, materiæ morbificæ exterminationem in ægri salutem omni opo molientis."—Sydenham, "*Observationes Medicæ*," §1. Quis circulus in probando !

" Helmetius, et, parum ab eo discedens, Campanella crediderunt febrem non esse morbum, sed morbi remedium, * * * ut peccantem materiem humoribus confusam eliminaret."—Baglivi, *Prax. Med.*, p. 72.

42. Page 28, Line 11.

As one illustration of the little estimation in which even now such matters are held, it may be stated that, winter before last, of a class of more than two hundred students invited by the author to attend a free course of lessons in Art-Anatomy by a competent teacher, with living models, less than twenty-five ever made their appearance, and only three or four continued through the course. One of these last has since had abundant reason to congratulate himself on his attention to these teachings.

43. Page 28, Line 24.

" It may now be affirmed that the practitioners of the present day are, speaking generally, almost as uninformed in this particular [the natural course and event of diseases] as were their predecessors fifty or a hundred years back."—Sir J. Forbes, " *Nature and Art*," p. 5.

44. Page 29, Line 8.

See " Origin of Species," by T. H. Huxley, F.R.S., 1863, pp. 91–100.

45. Page 30, Line 7.

" I remember to have been shown a manuscript copy of a New Practice of Physic, wherein the first article that catched my eye was on the scrofulous distemper, towards the end of which I perceived the word CURE in capital letters, followed by a number of recipes, which I immediately perused with the greatest eagerness, and then asked the author if he had known many instances of cures performed by those prescriptions. ' I never knew one in my life,' replied he ; 'but of what service would it be to describe a disease, if after the description I did not add the cure?' "—"*Medical Sketches*," by J. Moore, M.D., Lond., 1786, p. 64.

46. Page 30, Line 14.

It is not improbable that such a work from a fully competent hand may be given to the Profession before many years.

47. Page 30, Line 27.

See " Medical Communications of the Massachusetts Medical Society," 1863, p. 260.

48. Page 31, Line 11.

See " Practical Guide to the Study of the Diseases of the Eye," by Henry W. Williams, M.D., Boston, 1862, pp. 126–30.

49. Page 31, Line 19.

"Un signe infaillible qu'une science n'est pas constituée, c'est quand elle est encore une sorte de propriété commune. Mon portier n'hésitera pas à définir la maladie, à indiquer la cause, à prescrire le remède, et à prédire l'issue. Il s'en croit le droit; et il parait l'avoir, car on n'hésitera pas davantage à écouter son avis et souvent à le suivre."—"La Médecine et les Médecins," Paris, 1857, Vol. i. chap. i.

50. Page 31, Line 26.

See " Rational Medicine," by Jacob Bigelow, M.D., p. 29.

51. Page 31, Line 29.

" Even a moderate amount of knowledge of the general nature of diseases, and of the mode of operation and powers of the medical art, will make a man a better patient; make him more content with the treatment prescribed, be it energetic or inert; and make him repose greater confidence in his physician."—Sir J. Forbes, "Nature and Art," p. 14.

52. Page 33, Line 6.

" Ad plures migrabat." Sydenham, Ob. Med. i. 5, §61.

53. Page 33, Last Line.

Iliad, xxiv. 601, &c. Southeby's Translation, somewhat altered, and adapted.

www.ingramcontent.com/pod-product-compliance
Lightning Source LLC
Chambersburg PA
CBHW022028190326
41519CB00010B/1628